An Assessment of the Science and Technology Predictions in the Army's STAR21 Report

John Lyons, Richard Chait, and Jordan Willcox

Center for Technology and National Security Policy

National Defense University

July 2008

John W. Lyons is a Distinguished Research Fellow at the Center for Technology and National Security Policy (CTNSP), National Defense University. He was previously director of the Army Research Laboratory and director of the National Institute of Standards and Technology. Dr. Lyons received his PhD from Washington University. He holds a BA from Harvard.

Richard Chait is a Distinguished Research Fellow at CTNSP. He was previously Chief Scientist, Army Material Command, and Director, Army Research and Laboratory Management. Dr. Chait received his PhD in Solid State Science from Syracuse University and a BS degree from Rensselaer Polytechnic Institute.

Jordan Willcox is currently serving as a Research Assistant at CTNSP. He received his B.A. in International Relations from Connecticut College and is currently pursuing an M.A. in Security Studies from Georgetown University. Mr. Willcox has acquired previous experience in policy analysis, research support and database design from the Nixon Center and the Organization for American States.

Acknowledgements
In addition to the subject matter experts listed in appendix B, we acknowledge the cooperation and support of the Army Science and Technology Executive, Dr. Thomas Killion, and the Director of the Center for Technology and National Security Policy, Dr. Hans Binnendijk.

Defense & Technology Papers are published by the National Defense University Center for Technology and National Security Policy, Fort Lesley J. McNair, Washington, DC. CTNSP publications are available at http://www.ndu.edu/ctnsp/publications.html.

Contents

Executive Summary

This paper reviews the technology forecast assessments of the Strategic Technologies for the Army of the Twenty-First Century (STAR21) study conducted for the Army by the National Research Council in the early 1990s. The review in this paper was requested by the Army Chief Scientist, Dr. Tom Killion. The goal for STAR21 was "to assist the Army in improving its ability to incorporate advanced technologies into its weapons, equipment, and doctrine."[1] The objectives were to: identify the advanced technologies most likely to be important to ground warfare in the next century, suggest strategies for developing the full potential of these technologies, and project implications for force structure and strategy for the technology changes."[2]

This review was done by interviewing the Army's most senior scientists and engineers, who graded the technologies based on what has happened since 1992, when the STAR21 study was published. They considered which of the 104 topics in the technology forecasts were about right, which overestimated or underestimated progress, which were wrong, and what areas were missed entirely. The results were divided about evenly over four of the five categories. There were a very few topics judged to be underestimated; namely, distributed computer processing, computer memories, optoelectronics and photonics, development of biomarkers for medical diagnostics, functional materials, and manufacturing at the nano-scale. There were a few serious misses, such as significant developments since STAR21 that were not mentioned at all. The most significant change in technology that was missed was the development of the World Wide Web, which was enabled by the engineering of computer networks (the Internet) and the installation of fiber optic communication links to most parts of the world. About this time, the Netscape browser was introduced, which enabled PC users to maneuver around the Web and to communicate with any number of disparate systems. Another change that was missed was the extraordinary growth in the number of personal computers connected to the Internet.

The overall assessment by the scientists and engineers we interviewed is that the study earns a grade of about C. STAR21 did inform Army leadership of the future importance of these technologies. The fact that the study did not get everything right should in no sense argue against doing more such studies. Few technologists of the time could have predicted the impact of the explosion in information technology (IT).

The results depend on the approach taken and on the people doing the forecasting. The Air Force did a similar study in the 1990s, using its own scientific advisory board. It focused on the desired capabilities and then inferred from those the technologies of interest. The Army placed its focus on technology and reserved discussion of capabilities

---

[1] National Research Council, *STAR21 – Strategic Technologies for the Army of the Twenty-First Century* (Washington, DC: National Academy Press, 1992), iii. Available at http://books nap.edu/openbook.php?record_id=1888&page=R1.
[2] *STAR21 – Strategic Technologies for the Army of the Twenty-First Century*, iii.

in separate studies of systems. A Navy study at the National Research Council was intermediate between the Army and Air Force studies. (This paper briefly discusses the Navy and Air Force studies in appendix A.) In STAR21 it is possible to discern some biases arising from the participants. Biases should be excluded or balanced by the selection of the other committee members.

STAR21 increased the awareness of the impact of science and technology (S&T) on the Army's future capabilities at both the Army and Department of Defense (DOD) levels, as well as at the level of Congressional staff. It should be noted that S&T planning received a great deal of attention in the mid-1990s, just as it does today. The STAR21 study provided input to the Army's principal S&T planning document, the Army Science and Technology Master Plan. The present paper recommends that the Army should periodically sponsor similar forecasts of the future of S&T and makes suggestions as to how to improve the procedures for doing so. It also proposes that the three Services together sponsor these studies rather than doing them individually.

Introduction

Background

The urge to remain militarily strong has prompted U.S. defense planners to engage in technology forecasting. In particular, the analysis of emerging technologies leads to prudent DOD S&T investments. Among the preeminent examples of such analyses are studies undertaken by Theodore von Karman just after World War II.[3] The von Karmen reports represent an exhaustive review of S&T important to the military services. His forward-looking analysis projected the importance of unmanned aircraft, advanced propulsion, all-weather sensors, and target seeking missiles. Forecasting developments for the long term can provide useful context and perspective for decisions on the program content of the research and development (R&D) effort and the allocations of funds across these programs in the budgeting process. The reports are useful for increasing the awareness of senior leaders and those making decisions on the budget as well as demonstrating the significance of technology trends and their potential impact on the future of the institutions concerned. In the late 1980s, the Army initiated work with the National Academy of Science to conduct a study of predictions for technology in the next century, the Strategic Technologies for the Army of the 21st Century (STAR21).

The goal for the study was "to assist the Army in improving its ability to incorporate advanced technologies into its weapons, equipment, and doctrine."[4] The objectives were to: identify the advanced technologies most likely to be important to ground warfare in the next century, suggest strategies for developing the full potential of these technologies, and project implications for force structure and strategy for the technology changes."[5] This paper deals with the first objective, the technology identification and forecasts. STAR21, completed in 1992, was a multi-year and multi-volume effort. In a section termed Technology Forecast Assessments (TFAs) the study predicted developments in technology some 15 years ahead in the following S&T areas:

1. Computer Science, Artificial Intelligence, and Robotics

2. Electronics and Sensors

3. Optics, Photonics, and Sensors

4. Biotechnology and Biochemistry

5. Advanced Materials

6. Propulsion and Power

7. Advanced Manufacturing

8. Environment and Atmospheric Sciences

---

[3] Theodore von Karman, *Toward New Horizons* (Washington, DC: United States Army Air Force, 1945).
[4] *STAR21*, iii.
[5] Ibid.

The STAR21 study also provided, in a separate section, a Long-Term Forecast of Research (LTR) that extended out about 30 years into the future, and identified 11 major technological trends that cut across the traditional boundaries of scientific disciplines as follows:

1. The Information Explosion

2. Computer-Based Simulation and Visualization

3. Control of Nanoscale Processes

4. Chemical Synthesis by Design

5. Design Technology for Complex Heterogeneous Systems

6. Materials Design Through Computational Physics and Chemistry

7. Use of Hybrid Materials

8. Advanced Manufacturing and Design

9. Exploiting Relations Between Biomolecular Structure and Function

10. Applying Principles of Biological Information Processing

11. Environmental Protection

Also, several other long-term trends were noted in discipline-specific areas, such as: electronics, optics, and photonics; aeromechanics; molecular genetics; clinical medicine; atmospheric sciences; and terrain sciences.

Both the Navy and the Air Force conducted similar studies in technology forecasting. The Navy's study was conducted by the National Research Council's Naval Studies Board over about two years ending in 1997; [6] the Air Force study was completed in 1995.[7] Although all three reports discussed future capabilities and system concepts as well as technological advances, the Navy and Air Force reports placed greater emphasis on fully articulated capability goals, relative to pure technology growth. For example, the Air Force Scientific Advisory Board, in the Foreword to its report, describes it as "an integrated, capability-based, report."[8] Inevitably, some of the technological changes forecast overlap the STAR21 report: two obvious examples being unmanned vehicles and computer networking. In other cases–direct electric drive, advanced wind tunnel modeling, satellite materials–the services discussed distinctly separate technical challenges. At times, similar technological predictions led the Navy and/or Air Force to

---

[6] National Research Council Naval Studies Board, *Technology for the United States Navy and Marine Corps, 2000-2035* (Washington, DC: National Academies Press, 1997).

[7] The Air Force Scientific Advisory Board, *New World Vistas, Air and Space Power for the 21st Century* (Washington, DC: The Department of the Air Force, 1995).

[8] *Technology for the United States Navy and Marine Corps, 2000-2035*, iv.

create new and different system concepts for the use of those technologies. The topics covered in these reports are described, and comparisons with STAR 21 presented, in appendix A.

Approach

The present effort answers a query from Dr. Thomas Killion, the Army Chief Scientist, on the question of how well the STAR21 report made the TFA and LTR predictions. This report places the predictions in five categories: namely, the predictions were about right, underestimated progress, overestimated progress, were wrong, or missed some topics entirely. Our approach to the assessment relied on in-depth discussion with subject matter experts in each of the TFA and LTR areas listed above. For the trends listed in the section on LTR, we interviewed a group of 19 experts at the Army Research Office, and each expert commented on his or her area of specialization. Experts on the TFAs were drawn from many of the Army laboratories. We sought individuals that had positions within the Army S&T community with a rank of Senior/Technical (ST) or the equivalent. These individuals are considered to be national and international experts in their respective fields. Names and affiliations of the individuals contacted are provided in appendix B.

Of the interviews, two were conducted in person, the others by telephone or e-mail. The STAR21 study was published in two formats: one is a large volume of very detailed discussion, totaling about 675 pages, the other a summary document of the same material. The interviewees were given copies of the appropriate sections of the report; most were taken from the summary book. We found that in some cases where the interviewees said that a technological development was missed in the STAR21 study, the development was mentioned in another section of the study. Sometimes these items were in the separate STAR21 reports on systems, which were not focused on technology *per se*. We note this at the appropriate place in the text. Grades from A through F were assigned to each of the sections. Once this assessment was complete, it was possible to provide an overall grade to the study as a whole.

Summaries of the interviews, along with the graded assessment of each section, are presented in chapter 2. Chapter 3 presents an analysis and lists of the predictions by category: namely, were they about right, under- or overestimated, wrong, or missed. A more detailed list is provided in appendices C and D. This chapter also presents discussion and conclusions about STAR21; chapter 4 recommends more such studies, and offers specific suggestions for future Army or DOD studies of this kind.

Results of the Interviews of the Experts

The interviews in this paper covered two separate sections of the STAR21 study: first, the discussion of the LTR and second, the TFAs. The following presentation is in accord with this structure.

Long-Term Forecasts of Research

This portion of the present study, based on interviews with personnel from the Army Research Office, evaluates the 11 long-term research categories noted previously. **The overall grade for the LTR is C.**

1. ***The information explosion.*** (Drs. West, Arney) **The grade for this section is C.** This was a good explanation of needs, but it underestimated the problems, specifically of information saturation. Conversely, the likely solutions were overestimated. The Army is addressing all of these problems, but there has been less progress than predicted. The study missed the information security/assurance problem. There is no section on communications research; nothing on the internet and network science; and nothing on search engines.

   STAR21 also missed the potential use of biological principles to make information storage more intelligent—redundant, flexible, adaptable information storage that can learn, self-replicate, self-repair, and adapt according to the needs of the user.

2. ***Computer-based simulation and visualization.*** (Drs. Coyle, Skatrud) **The grade for this section is B.** This section did not give enough credit to mathematical aspects of simulation. The new multicore chips present a difficult challenge. There is not enough collaboration between physical scientists and computer people/numerical analysts. There has been good progress in modeling combustion and detonation reactions at the surfaces of explosives. This applies in general to modeling in chemistry. There is a need for more work in the engineering fields.

   The study missed the possibility that in *aeromechanics*, computer simulations on new supercomputer architectures combined with advances in computational fluid dynamics, composite structural dynamics, and aeroelasticity will contribute to the goal of complete *aerostructural simulation* (another example of trend 2) for both hover and steady level flight, as well as maneuver.

3. ***Control of Nanoscale Processes*** (Drs. Reynolds, Prater, Woolard, Gassner) **The grade for this section is B.** This is a good write-up, although it is mostly on electronics and misses other applications. The discussion of molecular integrated circuits is "iffy." The section did not emphasize quantum devices enough, control via light fields, and the use of metamaterials. The use of biomolecules as sensing devices is progressing from basic research to applied work with some potential systems in evaluation.

Modern computing and sensing technology are rapidly reaching a level of miniaturization and sensitivity at which inherent quantum uncertainties, due to indirect but continual measurements, can no longer be neglected. Due to dynamical analogs of the uncertainty principle, it is impossible to gain information about the state of a nanoscale scale quantum system without perturbing it in a manner that cannot be determined beforehand. Controlling the dynamics or measurement of quantum systems via the manipulation of external parameters is a most important phenomenon that lies at the heart of several fields including molecular chemistry, quantum information, and quantum computation. As a result, filtering based quantum control theory is a rapidly growing research field.

The area of supramolecular chemistry, in which people examine and tailor molecules so that they interact in a geometrically prescribed way with other molecules to form useful molecular assemblies, doesn't seem to have explicit coverage in the report. This is a major field of materials science at present.

4. ***Chemical Synthesis by Design***. (Drs. Lee, Shaw) **The grade for this section is B.** The study didn't go far enough. The Army is doing design this way today. The study missed the importance of force fields. The area needs new algorithms. Significant advances have been made; the field is moving into the life sciences. The study missed multi-functional materials. This section overlays trends 3 and 6. The section missed some of the complexity of such modeling.

5. ***Design Technology for Complex Heterogeneous Systems.*** (Drs. West, Chang, Mullins) **The grade for this section is B.** This section fits under network science and was a harbinger of the current programs on networks. The Army has recognized the challenge (credit goes to Dr. Killion's office) and has active research programs. The Army now knows a lot of things that don't work.

Control theory was missed. The Army is currently able to model non-linear systems only for the time variable and not for space variation. The study did not consider that modeling the design process is now out of favor. Mathematical studies of spatial-temporal nonlinear filtering for optimal controls and stabilization of these systems with incomplete observation/ information are urgently needed. Although it is desirable to develop feedback optimal control strategies, the inapplicability of dynamic programming principles in these cases, due to the non-Markovian nature of the systems, pose a great mathematical challenge for many years to come.

6. ***Materials Design through Computational Physics and Chemistry.*** (Drs. West, Chang, Mullins) **The grade for this section is C.** All of the trends 1-5, and 8 apply here. The section is too brief. Currently, scientists cannot model the processes to make materials; for example, molecular beam epitaxy (see Trend 2). Again the computer and physical science researchers need to collaborate more. There is a need to develop design rules based on computer models.

The discussion misses most of the techniques in use today, namely, molecular dynamics, classical Monte Carlo methods, quantum Monte Carlo methods, and statistical physics approaches.

More importantly, to truly design materials, the inverse problem not the "forward problem" must be solved. The traditional methods solve for the energy states and properties for a specified molecule or solid. The inverse problem begins with the property of interest (say conductivity, hardness, or transition temperature) and asks what material has the required property.

7. *Use of Hybrid Materials.* (Drs. Stepp, Prater) **The grade for this section is C.** The study understated the challenges. It has proved to be hard to obtain more than one function in a hybrid. The potential of smart structures was overstated, but the study nonetheless made an important statement as to the criticality of this concept. Smart structures remain a high-impact research topic.

8. *Advanced Manufacturing and Processing.* (Dr. Lampert) **The grade for this section is D.** This section is too short. It parallels the sections on materials design. There is nothing on the characterization of the processes. It deals largely with small scale (in lab) processes; scale-up of technology from the laboratory to the factory is not discussed. The section refers to biological processes, such as self-assembly and molecular recognition. There is not enough discussion of the physics and chemistry of, for example, manufacturing materials. The study does not mention bio-inspired or bio-directed nano-patterning and nano-assembly.

In the large book, the study discusses computer-aided manufacturing and making individual parts on demand at or near the front lines. Given the changed nature of the Army's challenges in the war on terror, this approach has not been emphasized. As presented, it does not appear to be a basic (6.1) research area.

9. *Exploiting Relations between Biomolecular Structure and Function.* (Drs. Stepp, Mattina, Kokoska) **The grade for this section is B.** The study makes an assertion that the principles are well enough understood to be used in designing materials. This is an overstatement. The predictions for vaccines and medicines are right. The study did not mention the key interactions between biomolecules and their environment. The field is very multi-disciplinary. The section missed biofuels, and missed the need for better understanding of injuries so that better protective gear may be provided. The study does not discuss bio-inspired materials such as armor based on clam shell structure and chemistry.

There is a list in the study of potential applications that relate to bioengineering. However, the most recent and important applications are missing: bio-templating of materials, bio-inspired armor, tissue regeneration, and biofuels.

10. *Applying Principles of Biological Information Processing.* (Dr. Kokoska) **The grade for this section is C.** The emphasis here was on large-scale biological

systems, such as neurological systems, and did not discuss other scales such as signal transduction at the molecular or cellular level in, for example, synapses. The study was correct on the potential for using what is learned of the biological basis for learning in training and in performance of information-intensive tasks.

Emphasis should be placed not just on larger scale processes (for example, pattern recognition, biological basis for learning and memory), but also processes at smaller scales, such as the molecular level (receptor binding/subsequent signal transduction) and the cellular level (modes of cell-cell communication in both bacterial and eukaryotic cells; electrical signaling at the neural synapse mediated by neuro- transmitters) and for multi-scalar processes. This field also should embrace the thinking and applications that are being developed in the fields of network science/systems biology to provide a more integrative framework for modeling and understanding these processes.

The Institute for Collaborative Biotechnology is working in this area at several levels. This section could be broadened beyond biological systems to include other kinds of networks.

11. ***Environmental Protection.*** (Drs. Preston, Buchholtz, Bach) **The grade for this section is B.** This area remains very important. The write-up is from the perspective of the warfighter; there is little on the situation impacting the general public. Since about the year 2000, DOD has required an analysis of the environmental costs of new programs on a life-cycle basis. This is an approach looking forward as opposed to looking backward; for example the remediation of the results of past practices. The study was correct that the Army has had to expend major effort to ameliorate past environmental damages. The discussion of hazardous wastes did not include depleted uranium. This is missing the human protection element and the poorly understood effects of depleted uranium, nanomaterials, and JP8 on human health. There was no discussion of the stockpile and non-stockpile problems of the demilitarization of chemical munitions—a major program for the Army. There is nothing on the effect of Army practices on the climate. Environmental concerns affect the sustainability of the field Army, for example in recycling or disposing of waste on the battlefield (see the write-up on deployable bioproduction facilities in Section 4 above on Biotechnology and Biochemistry).

Technology Forecast Assessments

The predictions for the seven sections under Technology Forecast Assessments (TFA are given below. **The overall assessment of the TFA is a grade of C.** Individual grades are found at the beginning of each section.

1. ***Computer Science, Artificial Intelligence, and Robotics*** (Dr. Lyons, Mr. Sciarretta) **The grade for this section is C.**

The study's opening discussion seems to be about right. It was correct about the challenges of developing integrated systems. The suggestion that it will be possible to design and make Very Large-Scale Integration (VLSI) chips in forward areas is wrong. The discussion of simulation and visualization is on target.

The study's discussion of development of a battle control language was wrong.

The forecast of networks on the battlefield was right. But the study missed associated areas that are now very prominent: the internet and World Wide Web, wireless communications, ubiquitous PCs, blogs, and search engines. Because of the web, soldiers are able to use these features from the forward areas. Soldiers are using blogs to exchange experiences on the battlefield in real time.

The study missed the problems of information overload and bandwidth limits (however, this is inherent in the statement on prioritization).

In distributed processing, the study underestimated the rate of progress.

The discussion of human-machine interfaces is good.

Using smart mines as the paradigm for the evolution of the robotics program was wrong.

The discussion of unmanned aerial vehicles (UAVs) is satisfactory but missed micro-UAVs. (Micro UAVs were discussed in the separate report on Airborne Systems. However, today's micro-UAVs are much smaller.) Walking and leaping robots, though developed, do not seem to be taking hold in the Army. The study missed the uses of robots for Improvised Explosive Device (IED) work and entering caves and buildings. It also missed the potential use of robots teaming with soldiers. The Future Combat Systems (FCS) project is planning to have robotic vehicles working alongside humans.

Neural nets have not taken hold, at least not in the Army.

In relation to high-performance computing, the study missed the emphasis on massively parallel systems and the budding use of multicore chips.

Knowledge-based or expert systems have not progressed as predicted. The predictions on the importance of natural languages and speech are satisfactory.

The study missed the issues of cyber security and information assurance, the change to counterterrorism and IEDs, and the challenges of asymmetric warfare.

The study did not address the Global Positioning System (GPS) in this section nor the current widespread use of inexpensive hand held GPS receivers. (However,

GPS is mentioned—not in the TFA, but in the section on Electronic Systems. We did not review the systems sections of STAR21.)

2. ***Electronics and Sensors.*** (Drs. Pellegrino and Armintharaj) **The grade for this section is B.**

Electronics: The prediction that silicon would remain the workhorse in electronics was correct. The study correctly pointed out developments in III/V technology, although it missed the use of InSb for very high-speed circuits. The predictions for SiC and diamond were overly optimistic. Both have narrow niches so far. The study noted that SiGe alloy is now used in all FET channels. The prediction of the use of thin film superconductors turned out to be wrong. The only use of superconductors is in Superconducting Quantum Interference Devices (SQUID).

The study was right about Monolithic Microwave Integrated Circuits. These circuits have proved to be very useful. In current systems hybrids of gallium arsenide with silicon are in use for digital processing. The study was wrong about vacuum transistors. There are no current applications for this technology in the Army. The prediction was correct that bioelectronics would be attractive; this is now a very active research area.

The study was too conservative about computer memories. The timing was all right, but the study underestimated the very large reductions in the cost of memory chips. The study overestimated the use of Application Specific Integrated Circuits (ASICs). They are expensive and have been largely replaced by Field Programmable Gate Array (FPGA) chips. The study overestimated the use of wafer-scale devices. They are very expensive because manufacturing is very costly for the small quantities now used.

The predictions for the speed of computer operations were on target.

Digital signal processors have performed as predicted. They are now operating at teraflop rates for parallel processing. They are not made from gallium arsenide, because silicon has kept pace. In regards to increased computing power from alternate materials and Reduced Instruction Set Computer (RISC) designs, the study was wrong. RISC architecture came and went. Si remains the material of choice.

Automated design has progressed well but could be better if designers would share their software. Software remains proprietary. The study missed signal circuits and design—a small miss. Neural net chips have not done as well as predicted. They are likely not used in UAVs and Unmanned Ground Vehicles (UGVs). There is some use of neural net algorithms on FPGAs.

Communications: The study missed the wireless revolution, the rise of ubiquitous computing, the internet, the IT world and social networking (blogging, Facebook, etc.) It also missed Second Life—the virtual reality world that is a web-based community using avatars (to represent the participant).

Sensors: The study predicted the use of radar on aircraft for imaging targets on the ground. In fact the use in Iraq is for persistent surveillance from slow moving UAVs. The current sensors are in visible and infrared (IR) cameras. Research is on multimode, networked sensors. There is a need for integration of signals.

The study conceived of ground-based radar as upward-looking for aircraft and missiles. This was almost certainly because the study did not foresee the change in ground warfare (and stabilization and reconstruction).

The Army does have low-cost-multimode ground-based radar for close in uses such as spotting IEDs. For hovering surveillance—for example, from UAVs—downloading data is a problem. The Army is working on partial downloading. In target recognition, there is still a difficult problem in interpreting images.

The study was right about the use of acoustic sensors, especially in arrays. It missed the networking of the battlefield and the cueing of other sensors systems from the acoustic sensors. Table 3-1 in the Summary Volume on acoustic array sensors is okay, as far as it goes. The study was wrong on the use of SQUIDs for detecting the magnetic signature of tanks. SQUIDs don't work over long distances.

The study was right on the following: light-weight phased array antennae that are electronically steered; terahertz devices, especially for spectroscopic applications; and teraflop computing.

This section missed the ubiquitous use of GPS devices both in the military and civil sectors. Miniaturization has led to new military applications for GPS. The Air Force Joint Direct Attack Munition (JDAM) bombs use a combined GPS / inertial guidance package. The Army has a research program on a Micromechanical Systems Inertial Measurement Unit (IMU) planned for completion in FY07. The work is developing an IMU "deeply integrated with a GPS military receiver."

3. ***Optics, Photonics, and Directed Energy.*** Drs. Ratches, Pellegrino, Strickland, Wachs) **The overall grade for this section is C.**

Optics and Photonics: The study undershot optoelectronics. Optical interconnects are now available. The study was wrong on acousto-optics. There have been substantial advances in photonics.

Additionally, the study did not anticipate the recent change in ground combat to include stabilization and reconstruction. Therefore, it missed the focus on the individual soldier and small groups of soldiers.

The study missed the following: uncooled imaging IR sensors, the monoblock laser concept for hand-held products, new diffraction optics for the individual soldier, sniffers for detecting trace explosives, missed dual band multicolor arrays, and flexible displays.

Sensor fusion has not yet happened; the use of Schottky barriers for IR devices has not progressed. The study was right on short wavelength IR; that is now a hot area. The study overestimated the potential for neural nets; there is not much use today. The study was wrong about smart focal plane arrays; there is not much going on.

In order to use Laser Detection and Ranging (LADAR) as the study predicts, the Army needs 3D images. Also, neodymium lasers have eye-safe problems not foreseen by the study.

Directed Energy: The Army quit working on free electron lasers (FELs) in the early 1990s. The study overestimated this; perhaps even was wrong. The Thomas Jefferson Laboratory National Accelerator Facility in Virginia is working on FELs for the Navy. The average power is 2 kW at varying wavelengths. The study was wrong on the development of ground- or space-based FEL systems for shooting down missiles. The development of compact accelerators has not succeeded.

4. ***Biotechnology and Biochemistry.*** (Drs. Valdes and Drapeau) **The overall grade for this section is B.**

Biotechnology: The STAR21 study's emphasis on the interdisciplinary nature of the biotechnology arena is right on target (see the Army's Institute on Collaborative Biotechnology, for example). Both the National Institute of Health and the National Science Foundation are now focused on interdisciplinary work.

The prediction on deployable bioproduction facilities is correct. There is a project underway to demonstrate a system to convert waste (garbage) into useful products by grinding up the waste, fermenting it, and separating out the alcohols. The residue is treated thermally to produce a composite fuel gas. The alcohols and the gas, mixed with 10 percent diesel fuel, are fed into generators to produce electricity. These are small units for very localized use.

Biosensors: The Army has deployed antibody-based sensors as well as some based on polymerase chain reaction (PCR). Both are targeted at specific threats; the Army needs sensors for unknowns. The Army needs sensors that are integrated into the soldier's gear and uniform such that he doesn't have to worry about taking care of them.

Human Immune System Enhancements: The immune resistance enhancement predicted has not happened. Difficult ethical and technical issues have not been solved.

Novel biomaterials: The development of spider silk is not progressing. There is an idea for an under-armor polymeric system that, while not stopping projectiles, could be useful in keeping debris out of wounds. Biocomposite armor has not happened. There are unresolved toxicity issues with nanomaterials. No real applications yet have been found. The Army is still using standard mission-oriented protective posture (MOPP) gear. Bio-derived camouflage has not developed.

Medical diagnostics: These are limited in the field to preclinical tests before symptoms appear. The development of biomarkers for detecting brain hemorrhage is an active program area. There are projects on genomic markers for past exposure to chemical agents. The STAR21 study is conservative in this area.

Human Performance Enhancements: An example is the work on prostheses connected to nerve clusters—a DARPA/Walter Reed program is moving well. Work on load-bearing exoskeletons for individual soldiers is not progressing well. One issue is getting the power to operate such a device. In FCS, the solution is a robotic logistics carrier at the soldier or squad level. Options such as blood and gene doping are illegal. There is research ongoing for enhancing learning abilities and improving training. The Army is drawing on technology from video games. Sleep and sleep deprivation work are active research areas in the neurosciences. The interviewees mentioned enhancing metabolic efficiency in mitochondria.

5. *Advanced Materials.* (Drs. McNight, McCauley, Dowding, Fong, Gassner) **The overall grade for this section is B.**

Functional Materials: The section on Materials does not explicitly discuss the category of functional materials, that is, materials that perform functions other than structural. Examples are semiconductors, liquid crystals, materials for optics including lasers, sensors, magnetic materials, and so on. Many of these were covered in separate sections on electronics, sensors, optics and photonics; however, the topic deserved discussion in the materials section of the report. Developments such as organic light-emitting diodes and nano-clay reinforced polymer films were not anticipated. Another example is progress in solar energy conversion. The efficiency of solar cells has increased dramatically since STAR21 was written and promises to continue to increase. There are light-weight flexible solar blankets. Amorphous silicon is now being used in several applications.

Research on "smart" textiles is progressing. The idea is to weave sensor fibers into ordinary cloth used, for example, in soldier uniforms. This technique could lead to uniforms that monitor soldier physiology. The technique is also valuable in "smart" composites.

Ceramics: The report placed a great deal of emphasis on sustained high-temperature synthesis (SHS) (reactive ceramics) for processing armor materials. Primarily due to costs, this technique did not take hold.

Materials by design: The report missed the gap in capability in the intermediate zone between continuum mechanics and quantum calculations/molecular dynamics.

Metals: Metal matrix composites have developed more slowly than the report predicted. Ti/SiC fiber was a great material. A plant was built but shut down for lack of sales at the end of the Cold War.

Tantalum technology developed problems with texture. The discussion of modeling penetration is very good. The section on depleted uranium (DU) correctly predicted events. An alloy of DU and niobium has been developed. Research on heavy tungsten alloys was just under way in the Army when the report was written; but after the theory of adiabatic shear was developed, the work was redirected. Thus, this effort was overestimated. Nanomaterials were missed; work is ongoing. Regarding light metals, Lithium aluminum alloys are now commercial, but the Army is not using them. The Marines are using such metals in above-water parts of their Advanced Amphibious Assault Vehicles. Regarding magnesium, the Army is just getting up to speed. Al: composites of Al are just coming in.

Bulk amorphous materials by rapid quenching were missed in the report. (This refers to making pure amorphous metals by pouring molten metal on spinning cooled wheels for rapid solidification.) Generally, the study addressed the right topics, but development has been slower than projected.

Polymers: The discussion of organic matrix composites is very good. The same is true for resin matrix composites. Some are commercial. An especially relevant example is the sabot for the M829A3 kinetic energy round.

Some technology for rigid rod polymers has been developed, and used in products such as Kelvar, but these are expensive. The demand for high-performance carbon-based fibers dropped after the Cold War.

Materials manufacturing was missed. Forward-basing of parts manufacture predicted in the Manufacturing section has not happened. Smart manufacturing is becoming more sophisticated. This section missed nano-technology.

Use of High Performance Computing in Design of Materials: The Army has done design of energetic materials from quantum mechanics and molecular dynamics.

It also has done well at the continuum level. Modeling at the mid-scale or grain sizes is now a research topic. In regards to technology demonstrations: the Army does have accurate predictions of performance in demonstrations. These are not separated but rather incorporated within systems demonstrations.

Energetic Materials: The prediction regarding the nitrated cubanes was an overestimation. The improvements to capture the effects of the predicted 30 percent increase in energy content have not occurred. There have turned out to be other important variables. The notion of layered materials—sandwiches of explosives with inorganics—has not been used. Insensitivity is being achieved by controlling particle size. The study missed nanomaterials; this is an active research area.

Warhead Materials: The discussion of tantalum warheads for explosively formed projectiles (EFP) was about right. For shaped charges, tantalum is not as good as copper; this shows that the effect is not only from density. Substituting niobium for some of the tantalum in EFPs showed a small improvement, but is not worth the cost.

The idea of orienting crystals works for kinetic energy rounds but not for shaped charge or explosively formed penetrators.

The latest work on DU alloys is approaching the performance of unalloyed DU. However, the work has stopped.

Work on tungsten alloys was pushed by Congressional add-ons, but is not going well. This research area was overestimated. Applications of tungsten in penetrators against light armors have not worked.

6. ***Propulsion and Power.*** (Dr. Bryzik, Mr. Scharf, and Drs. Strickland, Wachs, Schmidt, Walker, and Mikkelesen) **The overall grade for this section is C.**

Power plants for ground vehicles: This section of the report received a favorable review. Diesel turbo compounding (TC) is now commercial; it raises the engine efficiency about 10 points—from 40 to 50 percent or so. FCS may use TC in vehicles. The integrated propulsion system (IPS) approach is going to happen—the report was right on. The Army might convert diesel exhaust heat to run electric generators (homopolar) rather than redirect it back to the diesel engine. This would be another version of hybrid power design. The all-electric and all turbine engines are not in the cards. The fuel economy is not good enough and the turbine requires too much air.

Hybrid power systems were first used in commercial systems, a use driven by pollution pressures. The exact role for hybrid power for FCS has not yet been decided. STAR21 did not describe the need for an architecture for power management on the battlefield.

High-Power Directed Energy: Regarding ionic solid state lasers, the goal was 100kW lasers, truck-mounted for defense against ballistic missiles. The new threats are close in from rockets, artillery rounds and mortars (RAM) as in Iraq.

A program called HILL at Defense Advanced Research Projects Agency (DARPA) has achieved 25kW and the hope is to reach 100kW continuous wave. STAR21 under-estimated this technology. One approach is coherent coupling of a series of lasers—up to eight. The study's discussion of coherent coupling was correct.

For ionic solid state laser arrays, the Army has gone from single crystal neodymium: yttrium aluminum garnet, (NdYAG) to ceramic slabs (also of NdYAG). These are made by high-impact pressing (HIP) powders; they are not single crystals. The HIP powders offer better doping control, better heat resistance, and less cross-polarization. The efficiency of these slab lasers will be higher than the single crystal lasers (20 percent vs. 10–15 percent) calculated as laser power out and electric power in from the wall plug. These powders are hard to make in terms of controlling particle size. There is a firm in Japan that can make them. The study got this about right.

Diode lasers for pumping are coming along well. The STAR21 prediction of efficiencies of around 50 percent is a little low but on the mark. Efficiencies may reach 70 percent. These have not yet been fielded.

The prediction of laser tunability from IR to ultraviolet (UV) in chemical lasers is not likely to happen. The study missed fiber optic lasers, an active research program today. Phase conjugation for laser beam correction in the atmosphere didn't work in slab-based lasers. The concept was good, but the project was dropped.

Radio Frequency (RF) systems such as high-power microwaves (HPM) are now focused on short range non-lethal use for active denial. Earlier emphasis was on more damaging power levels. There is now miniaturization technology for power supplies such that one can put HPM in an artillery shell.

Propulsion of Projectiles from Guns or Tubes: The liquid propellant technology was abandoned. Electro-thermo-chemical gun technology was dropped for FCS because the electric igniter requires too much pulsed power to be practical on vehicles.

The electromagnetic gun is still about 20 years away. Recent (since 1998) developments include improvements in design and predicted performance of the compulsator. The predictions for a small machine—500 kilojoules—were demonstrated in 1997. Problem is to reduce the size with higher output.

U.S. Army Armament Research, Development, and Engineering Center (ARDEC) is building a two megajoule machine with two counter-rotating compulsators, but it is too big to put on a combat vehicle. ARDEC needs 1.5 to 2 times more strength in the bands holding the compulsator together. Army research Laboratory (ARL) is working on composite bands based on graphite fibers. Downsizing the compulsator is the biggest challenge.

The rails and armature problems look more tractable. ARDEC has used coated rails to fire 10 or so shots at 2.9km/sec. The program has built a composite gun at 40mm; it looks promising. The plan would be to cantilever the gun on the platform just as the 120mm gun is on the Abrams.

The STAR21 Army plans separate demonstrations of the components during 2008. ARDEC plans an artillery demonstration by 2018–2020. If successful, fielding for artillery may come by 2025.

Missile Propulsion: Regarding gel propellants, STAR21 stressed sensitivity and signature management rather than improving performance. Changing the propellant to eliminate chlorine is likely irrelevant, except perhaps in dealing with secondary smoke. No gel propellants have been fielded; the topic is still in research. The section is judged to be too optimistic.

There is a need for energy management in start/stop cycles or in proportional throttling and maneuverability. This topic has not gotten enough attention. This is a strong driver for performance for both gel propellants and air-breather systems. Going to gels may increase the time constant in the wrong direction.

The study missed the possibility of using nanometals to help boost performance in gels. This would require some new physics as to the mixing of nanoparticles in fluid flows.

Contrary to the report, there has been no evolutionary improvement in solid propellants. No new systems have been developed.

Air-breathing missile propulsion systems are the only way to increase specific impulse. The Army should do more research in this area.

The reviewers could not verify the accuracy of the predictions on improvements in turbine efficiencies for Army engines. However, they did say that there have been considerable improvements in commercial engines.

The use of high-power beams to transmit energy from ground sources to flying UAVs was overestimated.

7. *Advanced Manufacturing.* (Drs. Gassner and McKnight) **The grade for this section is D**.

The discussion is predominantly on computer controlled processes—the combination of information systems to manufacturing. It predicts that parts for military systems will be made on such systems in the forward areas of combat zones. This has not happened.

This section was unusually brief and did not emphasize the broad applicability of micro and nano-fabrication, which continue to move forward in accord with Moore's Law. Also, the Army now can move things molecule by molecule or atom by atom using atomic force microscopes. However, self-assembly of designer molecules may be a better way of doing this for some applications.

8. *Environmental and Atmospheric Sciences.* (Dr. Bach) **The grade for this section is B.**

The Army is now trying to forecast out 2–6 hours and ranging to 25km. STAR21 had this about right. The boundary layer is 1–2km in height; events in this layer are important for their effects on Army operations.

Problems with newer weather forecasting models are being worked elsewhere, for example, the National Oceanic and Atmospheric Administration (NOAA) National Weather Service, the National Center for Atmospheric Research (NCAR), National Science Foundation, and U.S. Air Force; also there are international efforts on models. The Army will, in general, rely on civilian weather modeling. The Army works with the Air Force on the Integrated Meteorological System (IMETS). These models are good at small scale. Work is aimed at higher resolution at the regional scale. STAR21 was right about the smaller scale arena. Artificial intelligence has not flourished for weather forecasting.

Sensors for meteorology are not yet routine on UAVs. Atmospheric sensors should be flown on UAVs; the report was right about this.

For satellites, remote sensing LIDAR is at the prototype stage. Some radar systems have been launched but so far only for coverage in the equatorial regions (geostationary satellites). The geostationary satellites use visible and IR sensors. IR sensors show both the ground and clouds. Ground temperature is important. Ground-based radar is <u>N</u>ext-<u>G</u>eneration <u>R</u>adar (NEXRAD). The polar orbiters are in lower orbits and give data every six hours. Multispectral sensors are used from satellites for measuring soil moisture; there has been a steady increase in capability.

The section scarcely mentions GPS, and only in regards to identification of friend or foe by locating our forces on the ground.

In regards to weather modification, there are no DOD or Army efforts and not much Federal work. There is activity at some state agencies.

Transport and diffusion modeling for the atmosphere present difficult problems. The study was correct in stating the need for remote sensing. It was too optimistic on predicting progress.

Summary of the Assessments by Sections

Having presented the details of the expert reviews, Tables 1 and 2 summarize the results.

Table 1. Grades for the Sections under "Long-Term Forecast of Research"

Name of Section	Grade Received
The Information Explosion	C
Computer-based Simulation and Visualization	B
Control of Nanoscale Processes	B
Chemical Synthesis by Design	B
Design Technology for Complex Heterogeneous Systems	B
Materials Design through Computational Physics and Chemistry	C
Use of Hybrid Materials	C
Advanced Manufacturing and Processing	D
Exploiting Relations between Biomolecular Structure and Function	B
Applying Principles of Biological Information Processing	C
Environmental Protection	B
OVERALL GRADE	**C**

Table 2. Grades for the "Technology Forecast Assessments"

Name of Section	Grade Received
Computer Science, Artificial Intelligence, and Robotics	C
Electronics and Sensors	B
Optics, Photonics, and Directed Energy	C
Biotechnology and Biochemistry	B
Advanced Materials	B
Propulsion and Power	C
Advanced Manufacturing	D
Environmental and Atmospheric Sciences	B
OVERALL GRADE	**C**

For the 11 LTRs as well as the seven TFAs the grades average to C. **Thus, the assessment of the STAR21 study as a whole is C**. The experts evidently judged the write-ups and predictions to be about average. This is reflected in figures 1 through 5 in chapter 3, wherein there are many inaccurate or missed predictions. We note that the topics divide evenly among four of the five categories. The fifth—the underestimated topics—has only four entries.

Discussion

In more detail, figures 1 through 5 present the status of selected predictions in STAR21. In figure 1, the report rightly foresaw the advent of the Information Age. But as shown in figure 5, the study did not anticipate the combined effects of advances in fiber optic communications, computers—especially personal computers—the Internet, the World-Wide Web, and the wireless communications revolution. The study did predict the increase in speed of computers but underestimated the progress in computer memories (figure 2). The increase in capabilities of biosensors was predicted; this activity became a top national priority after 9/11. Several advances in materials science and engineering were correct; namely, the work on DU for armaments and armor, the development of organic and resin matrix composites, and improved design of solid state lasers. The estimate of the importance of diesel turbo compounding for ground vehicles occurred as predicted. All in all, our experts judged that about one quarter of the predictions (that were identified by our experts) were about right.

The study underestimated the progress and importance of photonics and optoelectronics. STAR21 also underestimated the successes in networking and distributed processing.

The study overestimated progress in about a quarter of the areas (see figure 3 for examples and appendix C for the full list). Coping with the large amounts of information coming from the networking of all segments of the battlefield (air and ground)—information saturation—is a problem in two respects. There are bandwidth problems, problems in prioritizing messages, competition between data messages and voice, and—most important—the overloading of the personnel on the receiving end. Interviews with returning soldiers from Iraq raise concerns about information overload. The study really did not anticipate such problems. The study authors thought sensor fusion would develop more quickly than it has; the same is true for automated target recognition (ATR). The Army can obtain the necessary images for ATR but still doesn't have a good way to interpret them.

Two gun programs have disappointed. The electro-thermo-chemical gun technology, which looked very promising in the 1990s and seemed assured of application in the FCS, has in fact been dropped from consideration for reasons related to the necessary power supply. The electromagnetic (EM) gun program continues along a very slow path. There are problems with the size and weight that make it unlikely to be fielded on a ground vehicle. The Navy has a substantial effort to place an EM gun aboard ships; they are test firing prototypes in 2008. Another long-standing problem is modeling the atmosphere so as to predict and track the movement of airborne toxins from chemical and biological munitions. Progress has been slow.

We list in figure 4 a few of the items about which the study was <u>wrong</u>. (In appendix C we list two dozen topics that our experts judged to be wrong.) The study forecasted the development of a special computer language for battle control. So far this has not made any progress. The study was wrong about modeling manufacturing processes. One expert commented that scientists do not know how to model a process such as molecular beam

19

epitaxy—and need to know. Various efforts to replace or enhance DU in warheads and armor have generally fallen short. Tantalum, tungsten, and niobium have been studied but are not used. The use of the free electron laser system has not materialized, although progress has been made at the Thomas Jefferson Laboratory, National Accelerator Facility.

Finally, we come to areas that our experts feel were <u>missed entirely</u>. We show in figure 5 a selection from the full list in appendix C, figure C-7. (The total of missed items in appendix C is just under a quarter of the overall total.) The most significant misses are in the fields of computers and communications. The concurrent development of the Internet, the World Wide Web, wireless communications, and fiber optic cabling has changed both military and civilian life. To these we can add the widespread use of search engines, the ubiquitous personal computer and the daily contacts between people and entities by means of instant messaging, blogs, and the like. Most if not all of this was not anticipated in STAR21. Oddly, the wide spread use of hand-held GPS receivers was not discussed in the TFA section. These have become essential on the battlefield and are widely used in civilian life. Robotics finally has come into its own for the military. On the ground the Army has robots for use in areas where the risk to life is considered unacceptable. Thus, the Army uses robots to handle IEDs and to enter caves thought to contain explosives. The military is working on smaller ground robots capable of exploring buildings. In the air, the UAV as surveillance vehicle is now commonplace; some UAVs now carry weapons. Mini- and micro-UAVs are in development for specialized applications. In advanced computing, massively parallel architectures now dominate high-end machines. New multi-core chips will further revolutionize computing by making the designs more compact. Biotechnology is another revolution that has produced genomics, bio fuels, new medical techniques, new agricultural products, and the promise of better control of disease. These developments likely will make treatment of combat casualties better and ease the recovery and life after combat for veterans.

Figure 1. Some Topics that Were About Right

- The information explosion and networking
- Human-machine interfaces
- Speed of computer operations; digital computer processors
- Light-weight phased array antennae—electronically steered terahertz devices, especially for spectroscopic applications and teraflop computing
- Biosensors
- Modeling penetration of armor; DU was correctly predicted
- Organic and resin matrix composites
- Diesel turbo compounding
- Ionic solid state laser arrays

Figure 2. Some Topics that were Underestimated – There has been better progress

- Distributed processing
- Computer memories

- Optoelectronics and photonics
- Functional materials
- Manufacturing at the nano-scale

Figure 3. Some Topics that were Overestimated—Progress slower than estimated

- The problem of information saturation
- Down-loading data from UAVs
- Automated or assisted target recognition
- Sensor fusion
- Metal matrix composites
- Electro-magnetic gun
- Electro-thermo-chemical gun
- Remote sensor LIDAR
- Transport and diffusion modeling for the atmosphere

Figure 4. Some Topics the Study got Wrong

- Modeling materials manufacturing processes
- Development of a battle control language
- Ground or space-based free electron lasers for destroying missiles
- Tungsten in penetrators for light armor

Figure 5. Some Topics the Study Missed

- Information security/assurance problems
- Biofuels, bio-inspired materials, tissue regeneration
- Smaller scales in applying principles of biological information processing
- Environmental threat from DU; demilitarization of chemical munitions
- Information overload; bandwidth limits on the battlefield
- Micro-UAVs
- Robots for IEDs and caves
- Usefulness of massively parallel computers
- Ubiquity of GPS receivers
- The wireless revolution
- The ubiquitous personal computers; the internet phenomenon; the virtual world
- Uncooled IR imagers; dual band multicolor arrays; flexible displays
- Fiber optic lasers

The accuracy of the forecasts in STAR21 varied considerably. In some cases the predictions were remarkably accurate; in other cases, they were either outright wrong or seriously under-estimated or overestimated technical progress. There were a few serious misses; for example, significant developments since STAR21 not mentioned at all. Predicting S&T 15 years into the future (about the elapsed time since STAR21 was written) was and is a very difficult assignment. It is remarkable that some things that were predicted to occur 20 to 30 years in the future have already matured into fielded products.

The most significant change in technology since the early 1990s has been the development of the World Wide Web, the installation of fiber optic communication links to most parts of the world, the Netscape browser, and the extraordinary growth in the number of personal computers connected to the Web. Today most library research and manuscript preparation is done on the PC. The computer and communications revolutions now enable collaboration in research to be conducted without regard to the location of the participants in real time. STAR21 was published before the significance of these developments became apparent. The next chapter considers the merits of conducting long-term technology forecasting and makes recommendations for how to go about it.

Concluding Remarks

The value of technology prediction depends on the interests of the intended reader. The National Research Council conducts what are termed "decadal studies" in various areas. Every 10 years or so the studies are repeated. The purpose is to have members of a community agree on priority areas for funding and present the results to Federal agencies, the Office of Management and Budget, and the Congress. Many of these NRC studies are stimulated by Congressional committees or by the concerns of the research communities sponsored by funding agencies, such as the NSF or the National Aeronautics and Space Agency. In the case of STAR21, the impetus appears to have come from inside the top management layer of the Army and was funded at the NRC solely by the Army.

STAR21 focused on S&T. In appendix A, two studies by the other Services are discussed. One study was done by the Air Force and published in 1995 and another by the Navy in 1997. The Air Force study discussed capabilities deemed necessary in the future, evaluating technologies as needed. The Navy study also does this in a number of areas but also includes a separate volume on technologies.

All of the experts we interviewed work in Army laboratories. They believe a study such as STAR21 is worth doing but not for the purpose of informing them and their Army colleagues. They believe themselves to be well aware of past and present trends in their fields.[9] In their opinion, such a study helps them to justify their program plans and budget proposals by informing Army leadership of the future importance of these technologies. The fact that the study did not get everything right should in no sense argue against doing more such studies.

In doing such a study one has to consider the scope (the number of areas of interest), the time horizon for the predictions, the time required to do the study, the selection of participants, the management of the study, and the structure of the report. STAR21 looked forward 30 years. This is very likely too far into the future. It is difficult in some

---

[9] John W. Lyons, et al., "Strengthening the Army R&D Program: A Strategy for Improving Army Research and Development Laboratories," *Defense and Technology Paper 12* (Washington, DC: Center for Technology and National Security Policy, 2005).

rapidly moving technologies to see much beyond five years. Predictions inevitably become shakier as the time of forecasts becomes longer. STAR21 was conceived in 1988 and finished in 1992. This, in itself, is a time span over which some of the topics most likely evolved and changed considerably. The scope of STAR21 encompassed a wide spectrum of S&T. It may have been easier to manage if the study had been subdivided, with different topics being studied in different years. STAR21 did not consider who would be developing the technologies; that is, universities, industry, government laboratories, or a combination of these. Neither did the study consider the impact of the rising technical competence of many more nation states than previously, nor the possibility of more collaborations both within the United States and internationally.

Different technologies advance at different rates. IT has progressed at mind-numbing speed, as have certain aspects of biotechnology. On the other hand, there are areas where advances are slower and incremental; for example, in the materials science of armor and armaments and in the aerodynamics of rotorcraft. Perhaps such differences should be taken into account in the organization of future studies.

The composition and balance of the committees has a bearing on the predictions. If a particular sub-discipline is over represented, or an individual of strong bias participates, one can see the impact on the predictions. If, at the time of writing one aspect of the discipline is dominant, then the report will reflect that. If the technology later moves in a new direction, the prediction will be off the mark.

The STAR21 reports could have been better organized. We found there to be considerable overlap of topics. This is not surprising, given the multidisciplinary nature of research. However, cross-referencing would help. The selection of topics to be discussed by a given committee did not always make sense. If the topics are not closely related, then a variety of expertise is needed; it becomes difficult to staff the working committee. We did not see the advantage to separating the sections on TFA from the discussion on LTR. We assume the latter was intended to deal with basic research (6.1) and the former, with applied work (6.2). However, the distinction is not clear. The separation required frequent repetition of discussion to no useful end. The use of a separate committee on the long-term trends seems, in retrospect, unnecessary. The STAR21 study took several years and was very expensive. In retrospect it appears that it needn't have taken so long. The Navy study took only about two years.

Recommendations

The Army should periodically sponsor reviews of critical areas in S&T to make predictions into the future. (We say sponsor, because having the study done by impartial outside committees provide more credibility with decision makers. This is as opposed to having the studies done by Army personnel.) The following are some suggestions as to how the studies should be done:

1. *The period between studies ought to be no more than 10 years.* A shorter interval would not allow enough time for trends to develop. A longer interval makes the context too uncertain for useful predictions.

2. *The study itself should not take more than a year and a half from the time the study is approved and the committee(s) established.* Usually NRC studies can be done and reviewed in about this time span. Sometimes they can be finished in a shorter time. This includes the well-known lengthy review and approval process at the NRC.

3. *The study should not cover all areas of technical interest at once.* The study should be broken into several parts, with each being a separate effort and probably done in a separate year. This arrangement would spread the cost over several years and would greatly ease the management of each component.

4. *The separate studies should encompass a logical set of technologies.* Where there is overlap between two or more technologies, these should be grouped under the same study committee.

5. *Fast-moving technologies should be studied together.* This can be done if the technologies are related. This is the case today for electronics, computers, and communications.

6. *Basic and applied aspects of a technology should be studied together and reported that way.* By combining the aspects of both S&T, the reader will have an easier time grasping the potential impact of the area.

7. *The membership of the study committees should be balanced among the various disciplines or sub-disciplines involved.* The composition of the study committee can have a major effect on the results. Inevitably, the committee report will reflect in some measure the opinions, often strongly held, of the members. Care should be taken to avoid allowing any one viewpoint or personality to dominate.

8. *In the future the three Services should consider doing such studies together as a single report of predictions of S&T.* Each service could then apply the results to their particular needs and capabilities.

Senior subject matter experts (SMEs) from the Army should be allowed to participate beyond liaison roles. One possibility is to require that some number of SMEs, perhaps drawn from the Army's STs or division chiefs in the laboratories, make presentations at the outset of the committee deliberations and be allowed to attend the meetings as observers and resource providers. (They should not, however, play any part in the committee's operations nor should participate in the writing of the report.) Future studies of technologies relevant to the Army will help inform decisions at upper management levels of the Army and stakeholders elsewhere in the government. Such studies will be useful in explaining, and arguing for, the resources needed to continue the S&T programs. The recommendations above are intended to make this possible.

Appendix A: Navy and Air Force Studies Comparison

Some years after the completion of the Army's STAR21 study, the Navy and the Air Force moved to release their own long-term forecasts of expected technological developments in their areas of operation. Although all three reports discussed future capabilities and system concepts as well as technological advances, the Navy and Air Force reports placed greater emphasis on fully articulated capability goals, relative to pure technology growth. Inevitably, some of the technological changes forecast overlap the STAR21 report: two obvious examples being unmanned vehicles and computer networking. In other cases—direct electric drive, advanced wind tunnel modeling, satellite materials—the services discussed distinctly separate technical challenges. At times, similar technological predictions led the Navy and/or Air Force to create new and different system concepts for the use of those technologies. This appendix will briefly review the processes followed by the Navy and Air Force, respectively, in creating the studies; discuss the topics covered and structure of the reports; and provide contrasts with the STAR21 report where appropriate. This appendix will generally not evaluate the accuracy of technical predictions or provide grades, though it may make mention of unusual concepts not covered by the Army.

Of the Armed Services' individual 'future forecasts' issued during the 1990s, the Navy Report was released the latest, in 1997, and may have benefitted from a greater opportunity than earlier reports to observe technological trends in the civilian world. The study, titled *Technology for the United States Navy and Marine Corps,* was itself a 10-year review of a 1986 Navy study conducted under Cold War circumstances but does not directly refer to 1986 predictions. In a similar manner to the Army's STAR21 study, the Navy organized the study under the authority and auspices of the NRC. The specialized volumes 2-9 of the study were written individually by established committees; a coordinative group was established to ensure proper division of topics and avoid duplication. The Navy helpfully included information on the organizational structure of the study in its summary volume, which discussed background and strategic topics beyond what is included in the specialized technological volumes. The study was completed in about one year, using well over 100 experts.

The Navy's summary volume contains a brief synopsis and a more expanded discussion of a series of major topics that roughly correspond to the following eight volumes. Like both other reports, the Navy adds complexity by using more than one conceptual map. Immediately before discussing the topical areas in the synopsis, the Navy presents an order of "assigning priorities," beginning with information tools, than personnel retention ("more effective use of people"), followed by weapons platforms, logistics, simulation, and finally R&D. There is then a discussion of the anticipated strategic environment and enemy force assumptions, followed by an introduction to each major topic of volumes 2-9—described in Table A-1—in more detail. The Navy Study, like the Air Force study, devotes significant space for "Information in Warfare," demonstrating the pervasiveness of the conceptual bundling of sensors, networked computers, satellites and high-speed virtual communications. Also of note are the volumes on "Undersea Warfare" and "Modeling." These titles demonstrate that—again like the Air Force—the Navy's research is organized around capabilities—or to put it another way, outputs, rather than discrete technological areas, per se—or to put it another way, inputs. These titles are also worth noting because of the disproportionate attention paid to these topics, relative to STAR21. The "Human Resources," "Weapons," "Logistics," and "Platforms" volumes, although perhaps containing occasional new technical topics or different system concepts, more frequently cover similar ground to the STAR21 study.

Table A-1: Comparison of Army and Navy Research Topics

Army TFAs	Navy Volume Titles
Computer & Robotics	Summary
Electronics/Sensors	Technology
Optics/Energy Weapons	Information in Warfare
Biotech & Chem	Human Resources
Advanced Manufacturing	Weapons
Advanced Materials	Platforms
Environment & Atmospheric	Undersea Warfare
Propulsion & Power	Logistics
N/A	Modeling & Simulation

Although most discussion in the Summary volume is organized along the above principles, the Summary Volume does present a clear and concise pair of tables at the beginning of Part II that briefly state key technological developments anticipated by the study in specific terms, followed by matching these technologies with the capabilities they enable. The second volume, Technology, essentially restates the bullet points in these tables, while explaining the technologies in question at greater length. The categories used in the technology volume are <u>very</u> similar to the STAR21 TFAs. "Materials," Environmental Technologies" and "Propulsion" are nearly identical, while "Computation," "Sensors," and "Automation" cover elements of STAR21 TFAs from Table 1. "Biotechnology & Chemistry" has been refocused under "Human Performance Technologies," and "Information & Communications Technology" has been given its own heading. "Enterprise Processes" retreats from some of the ambitious predictions of STAR21's "Advanced Manufacturing," instead discussing business-derived productivity software in the context of the Navy bureaucracy. Note that though the topics of the respective Navy volume and STAR21 TFA may be the same, the content frequently differs. For example, the Navy discusses direct electric-drive propulsion (one assumes with interest in its low acoustic signatures) at great length, whereas the STAR21 study devotes little or no attention to the topic.

Generally speaking, the successor volumes of the Navy Report do a good job of avoiding duplicative discussion of areas covered in the Technology volume. The technology volume is, like the STAR21 Report, intensely focused on fundamentals: for example, the "Sensors" section covers the beneficial and detrimental properties of silicon vs. other materials in electrical conductivity, as well as the implications for distributed sensors. In contrast, the "Sensors" section of Volume 3, "Information in Warfare" tends to discuss the platforms from which sensors will be able to operate, the metrics of what they will be able to see, and so on, sometimes naming previous topics, such as optical sensors and LIDAR but relying on assumptions as to the development path that were explained in more detail in the "Technology" Volume.

Similarly, the "Materials" section in Volume 2 discusses current and predicted general advancements in the ability to create computationally pre-simulated heterogeneous alloys, using thin-film techniques and nanoscale sensors that enable create adaptive materials. When materials are discussed in Volume 6, "Platforms", they are discussed in the context of specific limitations on engine performance and a need for metals with higher melting points. Thus, the Technology volume discusses trends in what could be called "global scientific growth," whereas the Platforms volume highlights specific warfighting needs.

One problem that can arise with this careful delineation between the technology predictions in Volume 2, and the systems projections in later volumes, is that it can occasionally be hard to determine where, or if, assumptions made in Volumes 3-9 about capabilities are grounded in technological breakthroughs in Volume 2. For example, the "Weapons" volume (No. 5) makes an explicit case that the Navy must develop or expand its capability for surface-to-surface precision guided missile strikes that will either replace or augment naval guns. Unfortunately, the discussion of this does not specifically address what, if any, technical progress from Volume 2 would be required to implement this vision. Some other notable differences include: the devotion of an entire volume to simulation (not only in terms of computing abilities but in process utilization, including acquisition); the practical discussion of Information in Warfare (Volume 3), which corresponds to the Air Force's "Information Applications" volume in *New World Vistas*, to challenges of reloading Vertical-Launch Systems on surface ships (Volume 8, Logistics), and beyond. The most similar sections are in discussions of computing and propulsion technologies and in UAV roles, but large swaths of the topical volumes discuss, often in great detail, technological challenges and capability needs not looked at by STAR21.

The U.S. Air Force's 1995 report, titled *New World Vistas: Air & Space Power for the 21st Century*, was completed at the end of 1995, a full three years later than the STAR21. The study was completed in just over one year from its authorization, about half the time of the STAR study. Whereas the Army chose to vest authority for the study with the NRC, the Air Force chose its own Scientific Advisory Board to conduct its study. The Scientific Advisory Board's more specific role and well-established relationship to the Air Force may be reflected in the content of the report, which focuses more on expected or desired Air Force systems and capabilities, and less on describing or predicting technological progress independent of platforms. Overall, the Air Force report often gives less of a broad picture at movement in the scientific universe. The level of effort made to explain the specific technical achievements that will enable the futuristic systems varies across the report's 15 volumes. As an example, a key section of the Munitions Volume is devoted to a "Self-Protection-Missile" that would use "reaction thrusters" to maneuver and shoot down incoming surface-to-air missiles. However, the volume does not discuss exactly what technological shortfalls prevent the creation of this missile at present, nor what advancements will enable it in the future, or what techniques will be used to achieve them. Similar choices of emphasis have been made in other volumes, especially ones whose titles imply capabilities, rather than technological areas: the Attack, Mobility, Space Applications, and Information Applications Volumes. In contrast, other volumes spend considerable time and effort discussing the technological underpinnings to the new systems they foresee.

As might be expected, the Air Force chose in some circumstances to investigate different areas of technological development than the Army. The table below provides a comparison of the 12 topics of the Air Force studies with the eight TFAs of STAR21. In addition to the topics discussed in each of the 12 volumes listed below and its summary volume, *New World Vistas* also included a classified volume and an 'ancillary' volume consisting of interviews, essays and speeches by members of the panel and other leading scientists, discussing broad visions of future technological achievements in the world and the Air Force.

Table A-2: Comparison of Army and Air Force Research Topics

ARMY	AIR FORCE
Similar or Identical Topics	
Propulsion & Power	Aircraft & Propulsion
Optics/Energy Weapons	Directed Energy
Biotechnology & Biochemistry	Human Systems & Biotech
Computer & Robotics	Information Technologies
	Information Applications
Advanced Materials	Materials
Dissimilar Topics	
Environment & Atmospheric	Attack
Advanced Manufacturing	Munitions
N/A	Space Applications
N/A	Space Technologies
N/A	Mobility

As was the case with the Navy report, there is considerable topical overlap between the Army and the Air Force studies, reflecting the manner in which areas of interest, such as Materials and Propulsion, cross service boundaries. However, the Air Force discusses some topics—namely, space technology and information warfare—at greater length than the STAR21 study. Notably, the "Information" volumes predict and incorporate the "information revolution," the internet, wireless networks, handheld GPS, tactical-level data sharing, ubiquitous integration of computers into complex machines, and autonomous agents to assist in interpretation. The study was conducted at a time when the first web browsers were already in public use. At times, the Air Force's focus on systems and roles leads some technologies to be discussed more than once. UAVs are discussed in several volumes—at least Aircraft & Propulsion, Mobility, and Attack—whereas the STAR21 study discusses all types of robotic vehicles in its "Computer & Robotics" assessment. Unlike STAR21, the *Vistas* summary volume does not systematically address the innovations discussed in the subsequent volumes. Like STAR21, it lays out a series of capability concepts: "Global Awareness," "Lethal & Sub-lethal Projection of Power," "Global Mobility in War & Peace," "Dynamic Planning & Execution Control," "Space Operations," and "People." Unlike STAR21, it subordinates its technical discussion under these concepts, discussing a few systems or technologies for each one. As the concepts do not line up exactly with the 13 follow-on volumes, it is left unclear if all the volumes are featured in the summary. Some parts of the summary seem to be new ideas; "Dynamic Planning & Control" extensively discusses a need for the Air Force to shrink mission planning times, a need not discussed in any of the topical volumes, although the IT used is featured elsewhere.

Meanwhile, the volumes themselves vary widely—a variation reflected in the titles of the volumes. For example, "Attack" clearly represents a function, rather than a technological research area, where as "Materials" represents a discrete area of research and application. The "function volumes" typically discuss new aircraft, capabilities, or programs without explaining what represents new technology, rather than new applications of old technology. The format of the volumes varies as well. For example, "Aircraft & Propulsion" separates needed technological developments from envisioned new systems, using a series of tables to capture the new technologies related to each new system, and then returning to another series of tables to collate all the previously noted new technologies by technical area. It goes as far as to grade the importance of each technology to each concept, a technique not seen elsewhere. Some volumes, such as "Directed Energy" and "Information Technology," take care to separate predictions about technological development from visions of USAF needs or future systems that will follow, while many other sections do not. The "Human Systems and Biotechnology" volume is split between a series of biotechnology 'visions' for the far future with no detailed discussion, and near-term discussions focusing on simulation and human resources software, rather like the "Simulations" and "Enterprise Applications" volumes for the Navy, although in less detail. Here and there entirely new concepts are presented—the GPS-guided, precision-glide airdrop logistical concept (in the "Mobility" volume), hypersonic wind tunnel modeling (Propulsion volume), triple-junction Gallium-Arsenide solar cells in satellite power sources (Space Technology), and so on.

Given the selective approach of the summary volume and the decision of many of the 13 panels to intermingle technological evolution with system concepts, and predictions with recommendations, it is beyond the scope of this appendix to systematically list all Air Force discussions that go beyond the STAR21 Report. Generally speaking, the materials, information, munitions, propulsion, and directed energy fields overlap substantially with STAR21. With the exception of the IT and space sections, most topics are covered in less detail than the STAR21 report, and a greater share of attention goes to setting goals for new research, rather than predicting independent advancement.

Appendix B. The Participants in the Interviews

Army Research Office (Long-Term Forecast of Research)

Dr. David Skatrud, Director, Army Research Office

Dr. Stephen Lee, Senior Scientist, Army Research Office

Dr. David Mann, Director, Physical Sciences Directorate

Dr. Peter Reynolds, Senior Scientist, Physical Sciences Directorate

Dr. Robert Shaw, Division Chief, Chemistry Division

Dr. Walter Buchholz, Program Manager, Life Sciences Division

Dr. Bob Kokoska, Program Manager, Life Sciences Division

Dr. Kurt Preston, Division Chief, Environmental Sciences Division

Dr. Walter Bach, Program Manager, Environmental Sciences Division

Dr. Dwight Woolard, Program Manager, Electronics Division

Dr. David Stepp, Division Chief, Materials Sciences Division

Dr. Bill Lampert, Program, Manager, Materials Sciences Division

Dr. William Mullins, Program Manager, Materials Sciences Division

Dr. John Prater, Program Manager, Materials Sciences Division

Dr. Bruce LaMattina, Program Manager, Mechanical Sciences Division

Dr. Bruce West, Senior Scientist, Math and Information Sciences Directorate

Dr. Chris Arney, Division Chief, Mathematics Division

Dr. Harry Chang, Program Manager, Mathematics Division

Dr. Mike Coyle, Program Manager, Mathematics Division

TECHNOLOGY FORECAST ASSESSMENTS

Army Research Office

Dr. Walter Bach – see the ARO list above. **Topic:** Environmental and Atmospheric Sciences

Army Research Laboratory

Dr. John Pellegrino, Director, Sensors and Electron Devices Directorate with Dr. Paul Armintharaj **Topic:** Electronics and Sensors

Dr. Stephen McKnight, Chief, Materials Research, Weapons and Materials Research Directorate (WMRD); Dr. James McCauley Senior Research Engineer (ST), WMRD; Dr. Robert Dowding **Topic:** Advanced Materials

Dr. Edward Schmidt, Senior Research Scientist (ST) Electromagnetic Gun Program Manager, WMRD **Topic:** Propulsion

Night Vision and Electronic Sensors Directorate (Communications and Electronics Research, Development, and Engineering Center)

Dr. James Ratches, Chief Scientist - Night Vision and Electrooptics (ST) **Topic:** Optics, Photonics, and Directed Energy

Armaments Research, Development, and Engineering Center

Dr. Richard Fong, Senior Research Scientist (ST) Warheads technology **Topic:** Energetic Materials, Advanced Materials

Tank and Automotive Research, Development and Engineeering Center

Dr. Walet Bryzik, Chief Scientist (ST) **Topic:** Propulsion

Missile Research, Development, and Engineering Center

Dr. Billy Walker, Senior Research Science (ST) , Dr. Clark Mikkelesen **Topic:** Propulsion

Space and Missile Defense Command

Dr. Brian Strickland, Chief Scientist (ST); Dr. John Wachs **Topic:** Directed Energy

Edgewood Chemical Biological Command

Dr. James Valdes, Senior Advisor in Biotechnology(ST) with Dr. Mark Drapeau, Center for Technology and National Security Policy) **Topic:** Biotechnology and Biochemistry

Center for Technology and National Security Policy, National Defense University

Dr. John Lyons, Mr. Al Sciarretta **Topic:** Computer Science, Artificial Intelligence, and Robotics

Appendix C: Categories of Assessments of Predictions

The lists here enumerate topics that the STAR21 study got about right (Figure C-1), underestimated (Figure C-2), overestimated (Figure C-3), had wrong (Figure C-4) and missed altogether (Figure C-5). The use of asterisks indicates the authors' judgment of the priority in importance to the Army. The use of double asterisks refers to topics judged to be of the highest priority; these are in the lists presented in chapter 3.

Figure **C-1. Topics that were About Right**

- The information explosion and networking **
- Progress in modeling for chemical and material synthesis *
- Vaccines and medicines *
- The emphasis on environmental remediation and avoiding future contamination *
- Importance of integrated systems
- Human-machine interfaces **
- Natural languages and speech technologies
- Silicon to continue; III-Vs important; MMICs; Bioelectronics *
- Speed of computer operations; digital computer processors **
- Automated design of integrated circuits
- Low-cost-multimode ground-based radar; use of acoustic sensors especially in arrays
- Light-weight phased array antennae electronically steered; terahertz devices, especially for spectroscopic applications; and teraflop computing **
- Short wavelength IR
- The interdisciplinary nature of biotechnology *
- Deployable bioproduction facilities
- Biosensors **
- Enhancing human performance
- Modeling penetration of armor; DU was correctly predicted **
- Organic and resin matrix composites **
- Design of energetic materials from quantum mechanics and molecular dynamics
- Tantalum EFPs—the ability for the jet to stretch as it flies
- Diesel turbo compounding **
- Integrated propulsion system (IPS) approach
- Coherent coupling of a series of lasers for directed energy *
- Ionic solid state laser arrays **
- Diode lasers for pumping
- Weather forecasting for 2-6 hours ranging to 25km—at higher resolution at the regional scale *
- Anti-materiel products (soft kill)

Figure C-2. Topics that were Underestimated—There has been better progress

- Distributed processing **
- Computer memories **

- Optoelectronics and photonics **
- Development of biomarkers for medical diagnostics *
- Functional materials **
- Manufacturing at the nano-scale **

Figure C-3. Topics that were Overestimated—Progress slower than estimated

- The problem of information saturation **
- Progress on quantum devices, control via light fields, and the use of metamaterials*
- Multifunctional materials
- Relations between biomolecular structure and function
- SiC and diamond electronics
- ASICs and RISC chips *
- Use of wafer-scale devices
- Ground based upward-looking radar
- Down-loading data from UAVs **
- Automated or assisted target recognition **
- Sensor fusion **
- Neural nets
- Free electron lasers
- Toxicity issues with nanomaterials
- Metal matrix composites **
- Tantalum and tungsten for armaments
- High performance carbon-based fibers
- Nitrated cubanes
- Phase conjugation for laser beam correction in the atmosphere
- Damaging levels of high-power RF energy
- Electro-magnetic gun **
- Electro-thermo-chemical gun **
- Enhanced performance of missile propulsion.
- Improved solid propellants
- Sensors for meteorology on UAVs
- Remote sensor LIDAR
- Transport and diffusion modeling for the atmosphere **

Figure C-4. Topics the Study got Wrong

- Modeling materials manufacturing processes **
- Manufacturing parts including VLSI chips on demand on the battlefield
- Development of a battle control language **
- Modeling the development of robotics on smart mines
- Walking and leaping robots for the Army
- Neural nets
- Use of thin-film superconductors
- Vacuum transistors

- Signal digital processors from GaAs
- RISC architecture *
- Use of radar from aircraft for ground imaging
- Smart focal plane arrays
- Ground or space-based free electron lasers for destroying missiles
- Development of compact accelerators
- Enhancing immune resistance
- Spider silk
- Biocomposite armor *
- Modeling failure in composites
- Layered explosives
- Tungsten in penetrators for light armor. **
- Tunability for chemical lasers to the UV
- High powered laser beams to transmit energy to aircraft or satellites
- Weather modification

Figure C-5. Topics the Study Missed

- Information security/assurance problems **
- The use of robots working in teams with soldiers
- Use of biological principles for information storage
- Multi-core chips *
- Importance of force fields in chemical modeling
- Control theory
- Importance to modeling of molecular dynamics, Monte Carlo theory, statistical physics
- Biofuels, bio-inspired materials, tissue regeneration **
- Smaller scales in applying principles of biological information processing
- Environmental threat from DU; demilitarization of chemical munitions **
- Information overload; bandwidth limits on the battlefield **
- Small and micro UAVs **
- Robots for IEDs and caves **
- Usefulness of massively parallel computers **
- Ubiquity of GPS receivers **
- The wireless revolution **
- The information age; ubiquitous personal computers; the internet phenomenon; the virtual world **
- Uncooled IR imagers; dual band multicolor arrays; flexible displays **
- Missed the gap in modeling materials between continuum mechanics and quantum mechanics/molecular dynamics
- Bulk amorphous materials by rapid quenching
- Nanomaterials for armaments
- Fiber optic lasers **